Amazing Plants

SCHOOL PUBLISHERS

Orlando   Austin   New York   San Diego   Toronto   London

Visit *The Learning Site!*
www.harcourtschool.com

The World's Biggest Tree

A giant sequoia named General Sherman is the largest tree (by volume) in the world. It grows in Sequoia National Park in California. It is also the largest living thing in the world. The General is as tall as a 27-story building—84 meters (275 ft)! It is 31 meters (102 ft) around at the base of the trunk. It would take about 20 people, holding hands, to reach all the way around the tree!

The total volume of the trunk is about 1,486 cubic meters (52,500 cubic ft). However, the volume is always changing because the tree is still growing. Each year, it produces at least 1 more cubic meter (36 cubic ft) of wood.

In 1978, a branch fell from the tree. That branch was more than 2 meters (almost 7 ft) around and more than 43 meters (141 ft) long. That one branch, by itself, was larger than any tree in the eastern United States!

The World's Largest Fruit

When you hear the word *fruit,* you probably think of an orange or some other tasty treat that you eat as a snack or a dessert. Plant scientists use the word *fruit* in a different way. To plant scientists, any part of a flowering plant that contains seeds is a fruit. By this definition, an orange is a fruit, and a pea pod, a squash, and a tomato are fruits, too. A pumpkin holds the world record for the largest fruit ever grown.

The world's largest tree:
General Sherman

GENERAL SHERMAN

Each year in the United States, there are contests to see who can grow the biggest pumpkin. Winners can get as much as $50,000.00.

Pumpkin growers start planning early in the spring. They buy the best seeds they can find. They start the pumpkin seeds indoors and then transplant them into rich soil. When pumpkins start to grow on the vines, the growers remove all but the one pumpkin that is growing fastest. Then all of the food that the leaves make can go to this one fruit. Growers give the roots plenty of water and plant nutrients. The fruit can gain 11 kilograms (24 lb) a day!

The biggest pumpkin so far was grown in Oregon in 2003. It weighed 628 kilograms (1,385 lb). That's big enough to make more than 1,000 pumpkin pies!

A pumpkin is the world's largest fruit.

The World's Largest Flower

The largest flower in the world is called rafflesia. It lives in the rain forests of Borneo and Sumatra. (See map on page 7.) This flower is about 1 meter (3 ft) across and weighs 11 kilograms (24 lb)!

Rafflesia is a very unusual plant. It has no leaves or stems. Instead, it grows inside a host plant. The host plant is a vine that is related to the grape. When a rafflesia seed lands on its host vine, it sprouts and grows tiny threads. The threads invade the host vine and take food and water from it. Rafflesia first appears outside its host plant as a small dark nub. After nine months, the nub grows into a cabbage-sized bud. Then the bud unfolds into a huge flower with five fleshy lobes. In a few days, the flower begins to smell like rotting meat. To humans, the flower really stinks. But to bluebottle flies, it smells like food! The flies land on the flower to eat. As they do so, they carry pollen from male flowers to female flowers. After pollination, the female flower produces up to 4 million seeds.

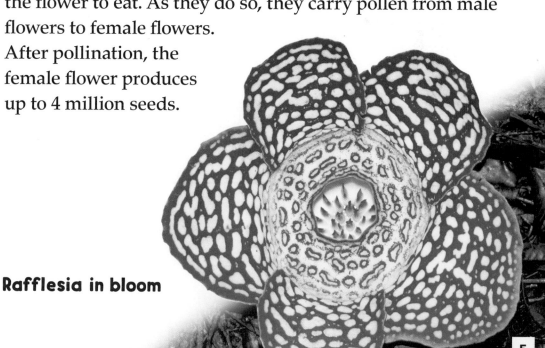

Rafflesia in bloom

The World's Largest Tuber

A tuber is a fleshy stem that grows under the ground. A potato is an example of a tuber. The largest tuber in the world can weigh as much as 75 kilograms (170 lb) or more! This tuber is the Titan Arum, also called the "corpse lily." It lives in the rain forests of Sumatra. (See map on page 7.)

A leaf stalk sprouts up from the tuber. The stalk can grow up to 6 meters (20 ft) high. One huge leaf looks like an umbrella and grows from the stalk. All of the food that the leaf makes is stored in the tuber. Then the leaf dies. After a rest period, the tuber can sprout again.

The Titan Arum lily produces the largest tuber in the world.

When the plant is mature, a huge, fleshy branch called a spadex can sprout from the tuber. This is a rare event. The spadex can grow 15 centimeters (6 in.) a day and can reach a height of 3 meters (10 ft)! At the base of the spadex are thousands of flowers with no petals.

When the corpse flower is in bloom, you can smell it from a half-mile away. It smells like—you guessed it—a dead body! The spadex smells the strongest at night, when carrion beetles come to the flowers to feast and pollinate.

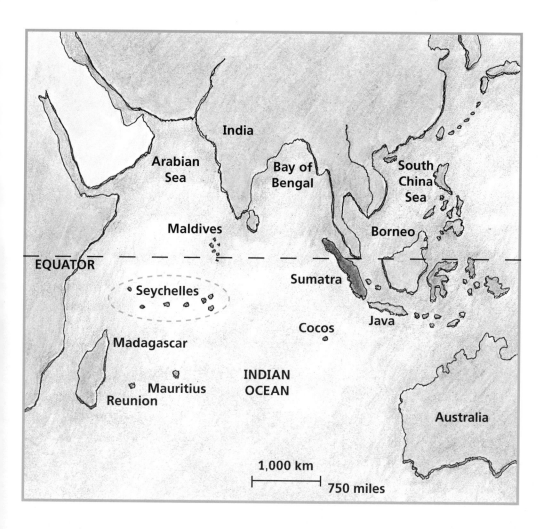

The World's Largest Seed

The largest seed in the world is from the giant fan palm. This palm tree lives in the Seychelles. (See map on page 7.) The seed is almost 60 centimeters (2 ft) across and weighs up to 20 kilograms (44 lb)!

During the 1700s, sailors used to find the giant seeds floating in the Indian Ocean. They called the seeds "coconuts of the sea." At that time, no one knew where the seeds actually came from.

The world's largest seed

People living in the Maldives thought they were from a mysterious tree that grew at the bottom of the sea. Anyone who found a giant seed was supposed to have good luck. Indian rajahs believed that the seeds could protect them from poisoning. They had goblets made from the seeds' shells and drank only from these goblets. The sultans of the Maldives kept this belief alive and sold the seeds to the Indian rajahs at very high prices. The seeds were almost worth their weight in gold!

The giant fan palm does not really grow at the bottom of the sea or protect anyone from poison, but it is quite amazing in other ways. The seeds float in the ocean for great distances and for a long time. When they wash up on shore, they can take root and grow. It takes a long time for the seed to sprout and grow. It sends a root down to a depth of 4 meters (13 ft). The first leaf doesn't appear until a year later, and sometimes not for much longer than that. The mature palm can reach a height of 30 meters (98 ft).

The World's Largest Leaf

A palm tree also holds the record for the world's largest leaf. It is a palm from western Africa called *Raphia regalis*. The leaves are up to 24 meters (80 ft) long! A single leaf is actually made up of several leaflets attached to a central midrib. In other words, the leaf looks a bit like a giant feather. The trunk of the tree is short, so the upright leaves reach from the ground to about eight stories high.

As you can imagine, the midribs of the leaves have to be very strong to hold leaves this large. Their strength makes the leaves a very good building material. Africans use the leaves for many purposes, such as building furniture, making baskets, and thatching roofs.

World's Oldest Living Plant

What is the world's oldest living plant? That depends on your definition of "living plant." Suppose that a plant gets old and falls over. The part that touches the ground sprouts new branches, and then the original plant dies. Are the original plant and the newly sprouted branches the same "living plant"? Plant scientists call the new branches "plant clones." This means that the original plant and its sprouts are genetically identical. In nature, many plants reproduce by cloning themselves in this way.

The oldest plant clones discovered thus far live in the wilds of Tasmania, an island off the southern coast of Australia. They are clones of a plant called King's Holly. They began growing more than 43,000 years ago! This was the time when modern humans were overtaking Neanderthals as the dominant species on Earth.

How did scientists find out that King's Holly is the oldest living plant clone? It is a real mystery story that began in 1937. A bushman and plant collector named Deny King discovered a strange plant while mining tin by hand in a remote part of Tasmania. Mr. King took some pieces of the plant to a plant scientist. The scientist had never seen the shiny-leaved plant before, but he guessed that it must be very old. The plant was named King's Holly in honor of the man who discovered it.

Tasmanian King's Holly: the world's oldest living plant

Over the years, scientists studied the plant further. They discovered that it reproduces only by cloning. It produces flowers, but it does not produce any living seeds. It simply gets old and falls down, and then it puts out new branches.

For many years, scientists searched for but could not find any other living specimens of King's Holly. Finally, on an expedition to the remote rain forest in southwestern Tasmania, scientists found about 500 remaining plant clones. These bushes straggle along a dark, damp gully in a line 1.2 kilometers (almost 1 mi) long. The climate in southwestern Tasmania is very cold, so cloning must happen very slowly. Scientists reason that a line of clones this long must have started growing several thousand years ago.

The scientists set out to look for fossils of King's Holly. Sure enough, about 8.5 kilometers (5.3 mi) away from the living plants, they found fossil leaves of King's Holly. The fossil leaves looked exactly like the leaves of living plants. Even the cells of the fossil plants were of exactly the same shape and size as the cells of modern plants. Scientists were quite sure that the fossil plants must be clones of King's Holly bushes living today. When scientists dated the fossils, they discovered the clones to be more than 43,000 years old. This finding made King's Holly the record holder for the oldest living plant clone yet discovered.